THE DESCRIPTION OF CREATION

THE UNKNOWN

The Description of Creation
By The Unknown

Edited by Vanessa Ta
Designed by Natalia Junqueira

ISBN 979-8-3789-0784-7
ISBN 979-8-3692-0096-4
ISBN 978-1-0879-3888-2

Copyright © 2023 by The Unknown

All rights reserved. No part of this book may be reproduced or used in any manner without written permission of the copyright owner except for the use of quotations in a book review. For more information, address: obscurity22@yahoo.com

First paperback edition May 2023

To those who want to be inspired.

There is nothing more honest
than the results of our actions.

In the beginning, there was nothing.

Nothing became so tedious

that eventually

NATURE began to show signs of boredom.

It grew more and more bored.

There was absolutely nothing

to SEE, HEAR, SMELL, TOUCH, or TASTE.

The ability to sense boredom

was all that existed.

There was nothing to do.

And there was nowhere to go.

BOREDOM took on a new meaning:

being completely confined.

The confinement was boredom within itself.

This state was untenable;

something had to happen.

The boredom eventually became so intense

that NATURE cultivated a force

to overcome the limitation.

The pressure of this force

grew to be stronger and stronger

until the force broke through

the confinement of boredom.

And NATURE's energy

was finally released.

Now, the energy could experience

the freedom of motion.

A whole new reality was ready to evolve

within the existence of this first microscopic placement.

As small as it was, it was enough to spawn

CREATION.

So, NATURE is the very first source of energy.

All it took was boredom

to reveal that nothing was not enough.

There had to be more

because nothingness is a waste of purpose.

And the purpose is to expand in every way possible.

If there is a way, NATURE will find it.

After the first stage,

it didn't take long for the process to continue.

Additional force made it easy for each step to lead to the next.

The further the ENERGY expanded,

the more space existed.

As it expanded

the space seemed to be empty.

Overall, each point formed a skeleton

to hold the entire space altogether.

And the emptiness evolved

into a whole new REALITY.

This new form needed purpose.

Allowing the nature of expansion to follow through,

it began to grow

in every direction to fill itself out,

creating an ILLUSION of endless space

that consumed itself completely.

And all the energy became trapped,

running circles within its form.

Increasing the speed to overcome the boundaries,

the energy's speed advanced to a state

where photons were created through radiation,

a process that allowed LIGHT to penetrate

through the darkness of space.

It left a trail that captured a path of colors.

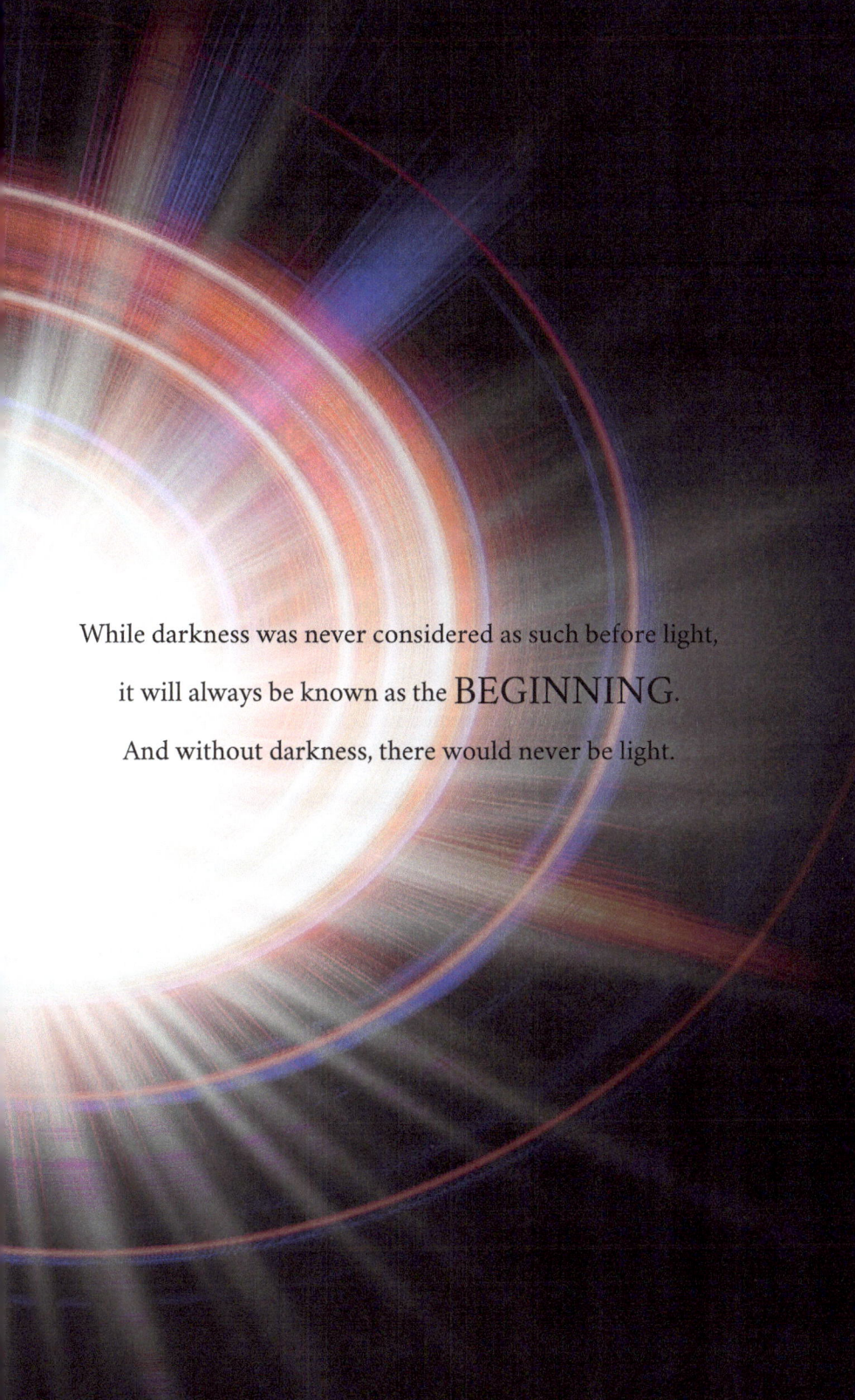

While darkness was never considered as such before light,

it will always be known as the BEGINNING.

And without darkness, there would never be light.

So, in the wake of a new REALITY,

the light filled the space,

while darkness disappeared from existence.

This became an overwhelming moment

within creation.

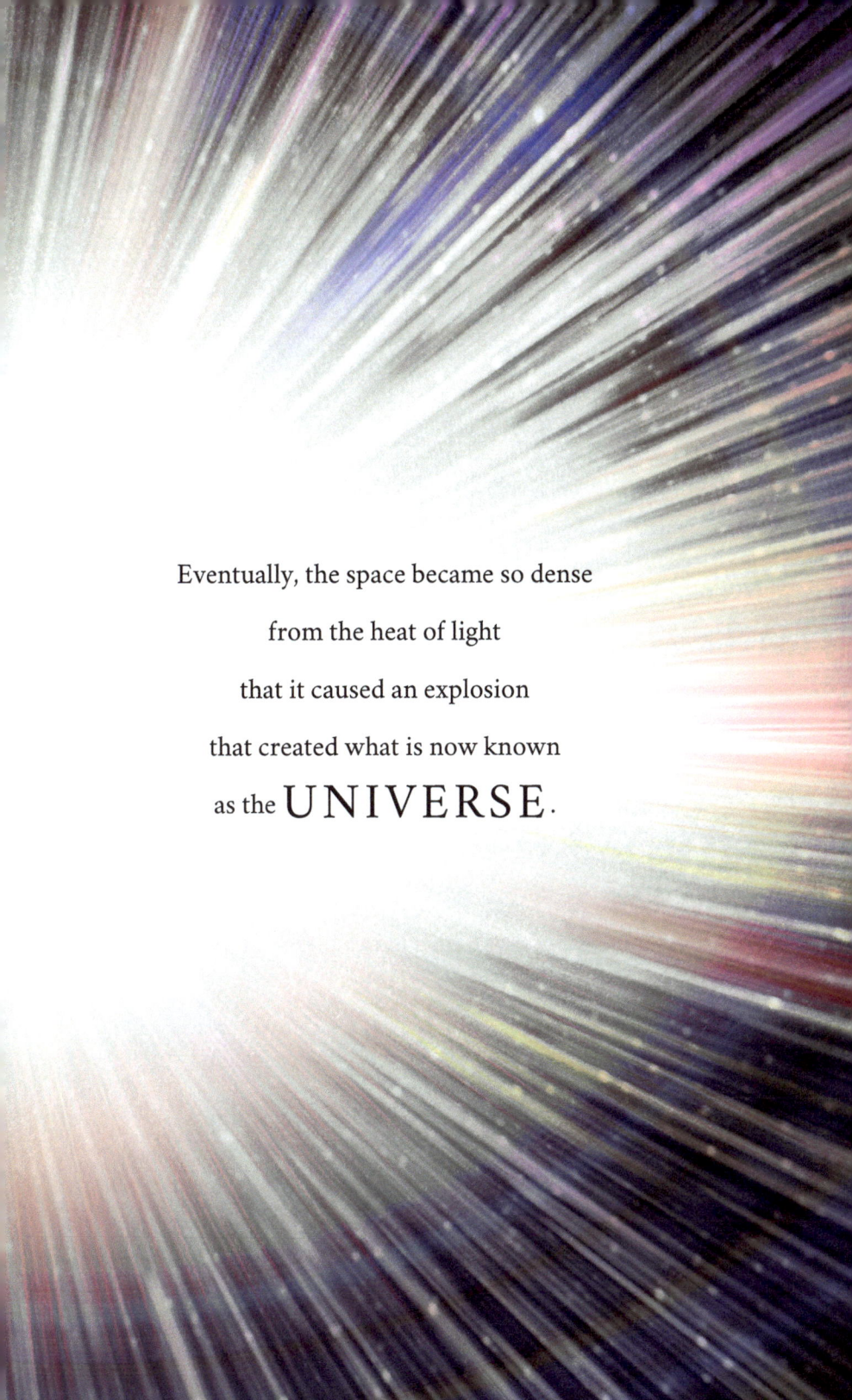

Eventually, the space became so dense

from the heat of light

that it caused an explosion

that created what is now known

as the UNIVERSE.

This was a defining moment in existence.

The ability to overcome

created the will to move forward,

because being able to make the impossible possible

gave PURPOSE altogether.

And the result changed everything.

Now, the point has been made.

There's only one question to ask.

What does it take to achieve complete satisfaction?

There's only one way to find out.

The creation of ENERGY was only the beginning-

and energy allows anything to be possible.

With that being said,

you have your whole life ahead of you.

NATURE was able to create energy out of nothing.

Just imagine what you can do

with everything NATURE has provided you.

Don't allow boredom to get in the way

of achieving the goals that give you purpose.

Overcome any challenge

and create a LIFE worth living

through your eyes alone.

No one can live it for you.

Now, go live it for YOURSELF.

MESSAGE OF THE AUTHOR

Believe in yourself. And follow the results of your actions, so you can make a difference in your life. The rest of the world will follow.

www.ingramcontent.com/pod-product-compliance
Lightning Source LLC
Chambersburg PA
CBHW041506010526
44118CB00001B/33